YOUR KNOWLEDGE HAS VALUE

- We will publish your bachelor's and master's thesis, essays and papers

- Your own eBook and book - sold worldwide in all relevant shops

- Earn money with each sale

Upload your text at www.GRIN.com and publish for free

Holger Skorupa

Outline and evaluation of the the expensive-tissue hypothesis proposed by Aiello/Wheeler (1995)

GRIN Publishing

Bibliographic information published by the German National Library:

The German National Library lists this publication in the National Bibliography;
detailed bibliographic data are available on the Internet at http://dnb.dnb.de .

Imprint:

Copyright © 2008 GRIN Verlag GmbH
Print and binding: Books on Demand GmbH, Norderstedt Germany
ISBN: 978-3-640-25373-9

Submitted by   Holger Skorupa

Date of submission   20th of October, 2008

Course Information   Essay 1: Outline and evaluate the expensive-tissue hypothesis proposed by Aiello/Wheeler (1995).

"Diet, Evolution and Culture", 200809-ALGY 387. Supervisor: Jessica Pearson, School of Archaeology, Classics and Egyptology

Status   Erasmus & Exchange Students Program

1. Outline and evaluate the expensive-tissue hypothesis proposed by Aiello/Wheeler (1995).

Hominid evolution has been a widely discussed aspect concerning its effects on ecological, physiological, and behavioral as well as reproductive, and metabolic development of humans and non-human primates for plenty of years. During the last decade, a huge amount of investigations regarding large brain size in humans have tried to figure out whether the observation, that humans have a relatively larger brain size than other primates or non-primate mammals, is based on either the correlated decrease of other tissues[i] or significant foraging strategies[ii], or even both.[iii] Although it is common, that an enlargement of the brain – being a high expensive metabolic organ[iv] – has to lead in dietary changes or that it was influenced by nutritional variations during the evolution of humans and non-human primates[v], there are different possibilities to explain this evolutionary progress. In 1995, Aiello/Wheeler published their expensive-tissue hypothesis[vi] regarding the relationship between large brain sizes in humans and high-quality diets. The authors attempt to confirm the parallel between the increase of hominid brains, the obviously correlated decrease of the gastrointestinal tract and dietary changes based on animal protein. They constitute the large brain sizes in human and non-human primates and its connection to nutritional varies to be one of the most significant prime releasers of brain evolution.[vii] The submitted essay portrays the main arguments of Aiello/Wheeler. By outlining the authors' intentions concerning the energy balance in humans and non-human primates, answering the linkage between the basal metabolic rate and the relative brain size of humans in the negative, and ascertaining both the lack of data based on human and non-human primates' evolutionary development, and the weakness of recent isolated studies, the essay will make an appraisal of the usefulness of the expensive-tissue hypothesis. Therefore, several papers of scientists facing the prime mover theories will be introduced.[viii] This pattern appears to be the best to indicate the advantages and weaknesses in the line of Aiello's/Wheeler's arguments. The essential question will be, how early humans are able to fulfil the high energetic costs of their large brains – in the following named as cost question.[ix]

The expensive-tissue hypothesis suggests an important and therefore essentially notable relation between the brain size, the relatively small size of gut of early humans in comparison to non-human primates and the hominid dietary habits.[x] The decrease of early human's gastrointestinal tract was probably influenced by embedding higher-quality food as well as the introduction of cooking to make food more digestible.[xi] Aiello/Wheeler argue, that the change of diet to a qualitatively higher food preparation could have been the most significant possibility on the subject of the enlargement of the brain size during hominid evolution. They point out, that mankind has to expend a larger sum of their energetic budget on brain metabolism.[xii] These increased demands are compensated by both the reduction of the gut size and an associated change of early *Homo's* dietary behaviour during the two major periods of brain enlargement. The appearance of the genus *Homo* meant the first dramatic development of hominid brain size two million years ago. It was followed by the outcoming of archaic *Homo sapiens* in the latter half of the Middle Pleistocene.[xiii] In addition to Aiello/Wheeler several ecological, biological, and archaeological scientists are claiming that there had to be much more aspects for the overall reduction from the externalization of organic functions during the evolution. While Fish/Lockwood suggest a variation of associated factors like social behaviour[xiv] and foraging strategy[xv], other critics of the expensive-tissue hypothesis predict assorted biological settings.[xvi] Although especially Aiello recognizes these notable conclusions in several articles[xvii], the authors underline the obvious problems of only little scientific verification on the theme and therefore ascertain the absences of adequate data.[xviii]

Additionally Aiello/Wheeler criticize the attempt of answering the cost question by comparing the enlargement of brain size with a decrease of the relative mass of other organs having relatively high mass-specific metabolic rates. They argue that these scrutinises result in arithmetic rather than metabolic research. However, Henneberg is highly critical of the expensive-tissue hypothesis in this mathematic case.[xix] On the other hand Fish/Lockwood prove Aiello's/Wheeler's observations by comparative studies[xx]. Therefore the expensive-tissue hypothesis seems to offer a large space of scientific controversies. The authors defence themselves by pointing out to attempt, how the metabolic cost of relatively large brains was possible without both an enlargement of the basal metabolic rate and the mass-decrease of hominid gut.[xxi] Emphasizing their arguments the authors portray the observations with an illustration.[xxii]

Furthermore Aiello/Wheeler name cows as animals having relatively large guts that are characterized by extensive and complex fermenting chambers like the gut.[xxiii] Regarding cows as non-human mammals being typical for large quantities of food of low assimilation the authors predict the dependence of gut size on both the bulk and the digestibility of food. Thus they recommend that diets characterized by smaller aggregates of high-digestible food entail smaller guts, a simple stomach, and long small paunches.[xxiv] The criticism of Armstrong[xxv], Falk[xxvi], and Wrangham/Holland Jones/Leighton[xxvii] of simply observing primate data – although a very short investigation of cows' gastrointestinal tract is given – without presenting differences between primates' encephalization and other taxonomic groups like birds or bats is faced by Aiello/Wheeler. They admit that such data would be very useful if adequate statistics would be provided which always has to be a significant factor of objective research.

However several scientists and scholars express their opinions about the first part of the expensive-tissue hypothesis by Aiello/Wheeler. Especially the authors' cost question seems to be an immediately discussed theme. Falk, for example, points out that the expensive-tissue hypothesis is likely to be in contrast to most "prime-mover" theories calling Aiello's/Wheeler's research as a "prime-releaser" theory.[xxviii] He supports the authors' physiological and anatomical study, although adding his "radiator" theory as a useful term concerning further research.[xxix] Holloway and Henneberg, by contrast, examine Aiello's/Wheeler's cost question critically. While Holloway names the complexity of the brain including plenty of behavioral functions that can affect all aspects of human's life[xxx], Henneberg criticizes the obvious reduction of the evolutionary development of *Homo* regarding only two periods of changes in body mass as done by Aiello/Wheeler.[xxxi] Therefore, both critics seem to be representatives of the "prime-mover" theory. Moreover, Henneberg faces much more examples showing the overall reduction from externalization of functions like the muscle decrease or the size of the viscera servicing the body during hominid evolution.[xxxii] Both Holloway and Henneberg come to the conclusion, that Aiello's/Wheeler' first part of the cost question has to be enlarged by examining why humans need larger brains and if a bigger acumen was actually better. By disclosing that point, they argue for the development of foraging strategies and social behaviour of early *Homo* being significant factors of the enlargement of the brain mass, too. However, Aiello/Wheeler[xxxiii] – and in further studies especially Aiello – constitute that it would be false to do remarkable research only on a high-quality diet as the most obvious prime releaser[xxxiv], they answer the criticism shown above in the negative by not affecting the expensive-tissue's main issue directly.

After explaining the relation between the brain size, the relatively small size of the gut, and the changes in food quality, Aiello/Wheeler provide a detailed attempt to investigate this connection within human and non-human primates. Several examples are given to follow the intention, that there is no obvious linkage between the relative basal metabolic rate and the development of hominid brain size during the evolution.[xxxv] Aiello/Wheeler use *Presbytis rubicunda* in comparison with *Presbytis cristatus*, for example, to contrast different diet qualities.[xxxvi] While *P. cristatus* is known to be aware of poor-quality food regarding their energies, *P. rubicunda* had a high-quality diet. Giving another instance the authors name the *Hylobatidae* by accenting *Hylobates lar's* feeding on fruits rather than on leaves with *Hylobates syndactylus* that is conscious to eat leaves rather than fruits.[xxxvii] However, these observations show a positive correlation between relative brain size and the size of the gastrointestinal tract of the *Old World Colobinae* instead of giving an evidence for a higher mass-specific metabolic rate. At that point of investigation Aiello/Wheeler criticize comparisons suggesting that *Old World Colobinae* were significant examples for having lower relative basal metabolic rates or guts with higher mass-specific metabolic rates indicating a larger energetic demand. Although both authors do not deny this theory, they refer to the weakness of such arguments. They point out the necessity of also observing several data such as brain size and the mass of the gastrointestinal tract to receive objective results. They follow the prediction of Ruff/Walker, here.[xxxviii] There are only few adequate data with reference to the metabolic rates of early human and non-human primates available.[xxxix] Furthermore, different individuals are investigated to receive necessary information. Therefore a strengthen comparison is very problematic and highly speculative. Apart from such criticisms[xl], Aiello/Wheeler present their observations by illustrating the relative brain mass and the relative gut mass in primates on the basis of both their observed and expected sizes.[xli]

The authors also predict, that their expensive-tissue hypothesis might give a new approach concerning anthropoid primates' relatively larger brains that were not linked to a high basal metabolic rate.[xlii] But – as an appreciable result of their research – they come to the conclusion that the significant parallel between the relative brain size and dietary habits of primates is a positive correlation between the relative brain size and the size of the gastrointestinal tract.[xliii] By illustrating the connections of a high-quality diet and an increased encephalization Aiello/Wheeler underline this aspect.[xliv] Besides, Fish/Lockwood have tested the authors' observation. They come to Aiello's/Wheeler's point of view and give their support to the expensive-tissue hypothesis, although it is added by Leonard/Robertson/Snodgrass, that several testings of differences in gastrointestinal size between primates and non-primate mammals have failed.[xlv] Nevertheless, the arguments of these scientists are not weak since it is clear, that humans have a relatively small gut size with reference to their body mass. But there are concerns remaining, however, asking how the decrease of the gut might help to balance the enlargement of energetic costs that became necessary by the increase of brain size during hominid evolution.[xlvi] Thus Aiello/Wheeler supplementary answer critics Henneberg[xlvii] in the negative by maintaining that it was definitely clear that the totally available energy was significant for surviving and hominid reproduction.[xlviii]

Additionally the expensive-tissue hypothesis' authors ascertain few arguments with less reputation. Being followed by the further examination of Leonard/Robertson/Snodgrass regarding the decrease of hominid tooth surface area in comparison to the *Australopithecines*[xlix], Aiello/Wheeler do not argue against other aspects which seem to be prime-releasers in the evolutionary development of human and non-human primates. They emphasize, however, that anatomical aspects like teeth or jaws do not necessarily have to be the only evidence of both dietary changes and the forthcoming of human body mass. Both scientists detect the coevolution of high-quality and easy to digest food such as animal protein, food preparation and the brain size.[i] By arguing in that case the reduction of the size of the gastrointestinal tract – and thereby the decrease of the considerable metabolic costs of the gut – is portrayed to be a possibly mostly significant issue to understand the hominid changes to higher-quality diets.[ii] Aiello/Wheeler therefore are highly critical of scrutinises that indicate the body of primates as an isolated construction. Such verifications examining the brain to explain the cost of encephalization, for example, have to fail because of not having tested their results by the use of several other tissues, they admit.[iii] Concerning their criticisms, it is absolutely common to achieve as much information, data, primary or secondary sources, and correlated testings as possible which ought to be the basis of every well-researched and objective investigation.

The expensive-tissue hypothesis proposed by Aiello/Wheeler suggests that organs such as the brain and the gastrointestinal tract are expensive tissues. In short, the authors emphasize the enlargement of the energetic costs of these tissues, the larger they are. In addition Aiello/Wheeler observe a correlation between the gut size and the diet quality of primates. As a main argument of their research, Aiello/Wheeler maintain the compatibility of relatively small guts with a high-quality diet and food being easy to digest – such as animal protein. The expensive tissue hypothesis consequently predicts the evolutionary development of the brain size to be one of the most significant prime-releasers concerning the consumption of greater quantities of animal products without ignoring the possibility of further factors that might have influenced the evolution of human and non-human primates. Aiello's/Wheeler's investigation therefore seems to be less speculative than most of the released prime-mover theories including tool production, language, warfare, and reproductive strategies and/or other aspects of early socialisation. However, these theories are also useful and consequently important to notice, but they could not be well tested. On the other hand, the expensive-tissue hypothesis uses representative and quantified data. Several papers show the advantages and disadvantages of the information given by Aiello/Wheeler by testing them in comparative biological – anatomical and physiological – observations. The authors' research allows both further studies regarding the correlation between the relative brain size, the size of the gastrointestinal tract and dietary habits of early *Homo*, and – in points of understanding the energy balance in human and non-human primates even more momentous – testings of various studies on the main issue of the hypothesis.

Meanwhile, several critics answer some arguments given by Aiello/Wheeler in the negative. Especially the authors' claim of the compatibility of small guts with an easy to digest food based on animal protein is hardly discussed. In fact it is common, that animal products as a part of food preparation have to be observed more differentiated. Although animal protein is high of metabolic energy, it is biologically proved, that the human body also needs a specific amount of fat to fulfil increased energetic requirements. Additionally, changes in food preparation, aspects of sharing food, the metamorphose towards hunter-gatherer societies, reproductive developments, and other foraging strategies do not have to be denied. They are obviously of a significant importance to understand the progress of early *Homo*. Modern prime-mover theories indicate several brain studies and correlated metabolic investigations to maintain the complexity of the brain in that case.

Subsequently the expensive-tissue hypothesis is very useful in the case of comparatively anatomical and physiological data allowing further and more detailed studies on the main issue of Aiello/Wheeler. By also disclosing the importance of social aspects during the evolution of the brain size and dietary habits of human and non-human primates, the authors portray a possibility to connect several theories that have looked as if there were no linkages between them for various years. Unfortunately, Aiello/Wheeler and especially Aiello began to discuss these fields of study years after they had published their observations. In conclusion, the expensive-tissue hypothesis offers new basis of research. On the other hand, it cannot predict other physiological factors being important with the reference to the hominid brain evolution. Thus, an interesting and noteworthy question remains: how easy could it be to change dietary habits from living on plants to hunting animals?

---

[i] For further studies see Chivers, D. J.; Hladik, C. M.: Diet and gut morphology in primates. In: Chivers, D. J.; Wood, B. A.; Bilsborough, A. (eds.): Food *acquisition and processing in primates*. New York 1984, pp. 213-230. See also Leonard, W. R.; Robertson, M. L.: Evolutionary perspectives on human nutrition. The influence of brain and body size on diet and metabolism. *American Journal of Human Biology* 6 (1994), pp. 77-88.

[ii] Ray, K.; Thomas, J.: In the kinship of cows. The social centrality of cattle in the earlier Neolithic of southern Britain. In: Parker Pearson, M. (ed.): *Food, Culture and Identity in the Neolithic and Early Bronze Age*. British Archaeological Reports International Series 1117 (2003), pp. 37-44, pp. 39 and 42. See Dunbar, R. I. M.: The Social Brain Hypothesis. *Evolutionary Anthropology* 6 (1998), pp. 178-190. Furthermore see Aiello, L. C.; Dunbar, R. I. M.: Neocortex size, group size, and the evolution of language in the hominids. *Current Anthropology* 34 (1993), pp. 184-193.

[iii] See Leonard, W. R.; Robertson, M. L.; Snodgrass, J. J.; Kuzawa, C. W.: Metabolic correlates of human brain evolution. *Comparative Biochemistry and Physiology Part A* 136 (2003), pp. 5-15. Furthermore see Aiello, L.C.; Key, C.: Energetic Consequences of Being a Homo erectus Female. *American Journal of Biology* 14 (2002), pp. 551-565. See also Aschoff, J.; Guenther, B.; Kramer, K.: Energiehaushalt und Temperaturregulation. Munich 1971.

[iv] Leonard; Robertson; Snodgrass; Kuzawa (2003), pp. 5 and 9. Dunbar (1998), p. 179. For one of the first studies concerning the basal metabolic rate and its comparison with species of various body mass see Kleiber, M.: The Fire of Life. New York 1961.

[v] Aiello; Key (2002), pp. 552 and 554. Fish, J. L.; Lockwood, C.A.: Dietary Constraints on Encephalization in Primates. *American Journal of Physical Anthropology* 120 (2003), pp. 171-181, p. 171. This point of view targets dietary changes to offset the correlated increase of energetic requirements because of the enlargement of the brain being a high-expensive metabolic tissue. Furthermore see Chivers, D. J.; Hladik, C. M.: Morphology of the gastrointestinal tract in Primates. Comparisons with other mammals in relation to diet. *Journal of Morphology* 166 (1980), pp. 337-386.

[vi] Aiello, L. C.; Wheeler, P.: The Expensive Tissue Hypothesis. The Brain and the Digestive System in Human and Primate Evolution. *Current Anthropology* 36 (1995), pp. 199-221. See also Aiello's second reply on the expensive-tissue hypothesis Aiello, L. C.: Brains and guts in human evolution. The expensive tissue hypothesis. *Brazilian Journal of Genetics* 20 (1997), pp. 141-148.

[vii] Aiello; Wheeler (1995), p. 207. Furthermore see the testings done by Fish; Lockwood (2003).

[viii] Leonard, W. R.; Robertson, M. L.; Snodgrass, J. J.: Effects of Brain Evolution on Human Nutrition and Metabolism. *Annual Review of Nutrition* 27 (2007a), pp. 311-327. In the following as Leonard; Robertson; Snodgrass (2007a). See also Leonard, W. R.; Robertson, M. L.; Snodgrass, J. J.: Energetic Models of Human Nutritional Evolution. In: Ungar, P. S. (ed.): *Evolution of the Human Diet. The Known, the Unknown, and the Unknowable.* New York 2007, pp. 344-359. In the following as Leonard; Robertson; Snodgrass (2007b). See also Mac Nab, B. K.; Eisenberg, J. F.: Brain Size and its Relation to the Rate of Metabolism in Mammals. *American Naturalist* 133 (1989), pp. 157-167. For short comments on the expensive-tissue hypothesis see Falk, D. in Aiello; Wheeler (1995), p. 212 and Henneberg, M. in Aiello; Wheeler (1995), pp. 212-213. See Holloway, R. L. in Aiello; Wheeler (1995), pp. 213-214 for the same issue.

[ix] Aiello; Wheeler (1995), p. 199. For further information see also Foley, R. A.; Lee, P. C.: Ecology and Energetics of Encephalization in hominid Evolution. *Philosophical Transactions of the Royal Society* 334 (1991), pp. 223-232.

[x] Aiello; Wheeler (1995), pp. 200-201.

[xi] Aiello; Wheeler (1995), p. 211.

[xii] Leonard; Robertson; Snodgrass (2007a), p. 311.

[xiii] Aiello; Wheeler (1995), p. 208. Furthermore see Aiello (1997), p. 145. See also Leonard; Robertson; Snodgrass (2007a), p. 317. Furthermore see illustration *Table 1: Observed and Predicted Basal Metabolic Rates for a 65-kg Human Compared with Other Primates and Eutherians (Placental Mammals)* given in the index. The Table's investigations show that BMRs of mature individuals are significant of primates and therefore notable for placental mammals as a whole. The illustration also suggests an obvious influence of sex or age on BMR.

[xiv] Fish; Lockwood (2003), p. 181. See also Aiello; Key (2002), pp. 562-563.

[xv] Leonard; Robertson; Snodgrass (2007a), p. 317.

[xvi] Henneberg, M. in Aiello; Wheeler (1995), p. 213. Holloway, R. L. in Aiello; Wheeler (1995), p. 214. Leonard; Robertson; Snodgrass; Kuzawa (2003), p. 5.

[xvii] Aiello; Key (2002), pp. 562-563.

[xviii] Aiello; Wheeler (1995), p. 219.

[xix] Henneberg, M. in Aiello; Wheeler (1995), p. 213.

[xx] See Fish; Lockwood (2003).

[xxi] Aiello; Wheeler (1995), pp. 216-217. Furthermore see Aiello (1997), pp. 142-143.

[xxii] See illustration *Table 2: Organ Mass and Metabolic Rate in Humans* given in the index. Table 2 shows several tissues having high energetic demands like the brain, heart, kidneys, liver and gastrointestinal tract. It also indicates the relatively small percentage of resting human skeletal muscle in correlation to the brain. Additionally this figure influences the observation that the resting human skeletal muscle contributes only 14.9 % of BMR.

[xxiii] Aiello; Wheeler (1995), p. 206.

[xxiv] Fish; Lockwood (2003), p. 172. For further studies see especially Chivers; Hladik (1980).

[xxv] Armstrong, E. in Aiello; Wheeler (1995), p. 211.

[xxvi] Falk, D. in Aiello; Wheeler (1995), p. 212.

[xxvii] Wrangham, R. W.; Holland Jones, J.; Leighton, M. in Aiello; Wheeler (1995), p. 216.

[xxviii] The so-called "prime-mover" theories suggest both the dietary change since the appearance of early Homo two million years ago and the correlated development of gut size were the mostly necessary aspects of hominid encephalization. Apart from that these examinations claim hunting, tool production, social intelligence, and language to be essential evidence for the evolutionary brain enlargement. To learn more about the "prime-mover" theories see Leonard; Robertson; Snodgrass (2007a), p. 317, for example. See also Aiello; Key (2002), pp. 562-563. In contrast Falk underlines the highly speculative aspects of such theories in his point of view. See Falk, D. in Aiello; Wheeler (1995), p. 212. To overview the scientific intentions on both the "prime-mover" and the "prime-releaser" theories see especially Conklin-Brittain, N. L.; Wrangham, R. W.; Smith C. C.: A Two-Stage Model of Increased Dietary Quality in Early Hominid Evolution: The Role of Fiber. In: Ungar, P. S.; Teaford, M. F. (eds.): *Human Diet. Its Origin and Evolution*. Westport 2002, pp. 61-76.

[xxix] Falk, D.: Brain evolution in Homo. The "radiator" theory. *Behavioral and Brain Sciences* 13 (1990), pp. 333-381.

[xxx] Holloway, R. L. in Aiello; Wheeler (1995), pp. 213-214.

[xxxi] Henneberg, M. in Aiello; Wheeler (1995), p. 213.

[xxxii] Henneberg, M. in Aiello; Wheeler (1995), p. 213.

[xxxiii] Aiello; Wheeler (1995), p. 219.

[xxxiv] Aiello (1997), p. 145.

[xxxv] Fish; Lockwood (2003), pp. 171-172. See also Mac Nab; Eisenberg (1989).

[xxxvi] Aiello; Wheeler (1995), p. 206.

[xxxvii] For further studies on Hylobatidae see especially Milton, K: Primate diets and gut morphology. Implications for hominid evolution. In: Harris, M; Boss, E. B. (eds.): *Food and evolution. Toward a theory of human food habits*. Philadelphia 1987, pp. 96-116.

xxxviii See Ruff, C. B.; Walker, A.: Body Size and Body Shape. In: Walker, A.; Leakey, R. E. (eds.): *The Nariokotome Homo erectus skeleton*. Cambridge 1993.

xxxix Aiello; Wheeler (1995), p. 206. See also Holloway, R. L.: Within-species brain-body weight variability. A reexamination of the Danish data and other primate species. *American Journal of Physical Anthropology* 53 (1980), pp. 109-121.

xl Fish; Lockwood (2003), p. 173.

xli See *Figure 1: Relative brain mass versus relative gut mass in primates* given in the index. The illustration suggests the significant relationship between both the relative brain size and the relative size of the gastrointestinal tract. Consequently, animals with relatively small brains seem to have relatively big gut sizes. By contrast, animals with relatively small gastrointestinal tracts have relatively large brains. Regarding these observations the table supports the expensive-tissue hypothesis.

xlii Aiello; Wheeler (1995), p. 206. Aiello (1997), p. 146.

xliii Aiello; Wheeler (1995), p. 210. For further information see also Foley; Lee (1991).

xliv See *Figure 2: Increased encephalization and high-quality diet based on animal protein* given in the index. The illustration points out the main argument of the expensive-tissue hypothesis. It underlines the parallel between relative brain size and diet being at first a relation between relative brain size and relative size of the gastrointestinal tract. This correlation is influenced significantly by the quality of the diet. Furthermore, it implicates, that a high-quality diet may also be influenced by factors being recognized as prime movers. See Aiello (1997), p. 145. Furthermore see Dunbar (1998), pp. 184 and 187. See also Leonard; Robertson; Snodgrass (2007a), pp. 317-318.

xlv Leonard; Robertson; Snodgrass (2007a), pp. 319-320. See also Leonard; Robertson; Snodgrass (2007b).

xlvi See Falk (1990). See also Dunbar (1998), pp. 180-181.

xlvii Henneberg, M. in Aiello; Wheeler (1995), p. 213.

xlviii Aiello; Key (2002), pp. 552 and 562.

xlix Leonard; Robertson; Snodgrass (2007a), pp. 319-320, especially table 2.

l Aiello; Wheeler (1995), p. 211. See also Leonard; Robertson; Snodgrass; Kuzawa (2003), p. 9. Furthermore see Kleiber (1961).

li Aiello (1997), p. 144.

lii Aiello; Wheeler (1995), p. 211. Aiello (1997), p. 147. See also Leonard; Robertson; Snodgrass; Kuzawa (2003), p. 13.

2. Index of Tables and Figures

Table 1: Observed and Predicted Basal Metabolic Rates for a 65-kg Human Compared with Other Primates and Eutherians (Placental Mammals).

| | | Predicted 65-kg Mammal BMR | | | |
| | | Other Primates ($3.36\ M^{0.75}$) | | Eutherians ($3.39\ M^{0.75}$) | |
Sex and Age	BMR (W)	W	Difference (%)	W	Difference (%)
Male					
18-30 years	80.914	76.916	+ 5.20	77.603	+ 4.25
30-60 years	78.391	76.916	+ 1.92	77.603	+ 1.02
Female					
18-30 years	70.208	76.916	- 8.72	77.603	- 9.53
30-60 years	66.528	76.916	- 13.51	77.603	- 14.27
All					
18-30 years	72.561	76.916	- 1.76	77.603	- 2.63
30-60 years	72.460	76.916	- 5.79	77.603	- 6.63

Source: Aiello, L. C.; Wheeler, P.: The Expensive Tissue Hypothesis. The Brain and the Digestive System in Human and Primate Evolution. Current Anthropology 36 (1995), pp. 199-221, 202.

Primary source of Aiello; Wheeler: Schofield, W. N.: Predicting basal metabolic rate. New standards and review of previous work. Human Nutrition: Clinical Nutrition Part C 39 (1985), pp. 5-41.

Table 2: Organ Mass and Metabolic Rate in Humans.

Organ	Organ Mass (kg)	% Body Mass	Mass-Specific Organ Metabolic Rate (W.Kg^{-1})	Total Organ Metabolic Rate (W)	% Total Body BMR
Brain	1.3	2.0	11.2	14.6	16.1
Heart	0.3	0.5	32.3	9.7	10.7
Kidney	0.3	0.5	23.3	7.0	7.7
Liver	1.4	2.2	12.2	17.1	18.9
Gastro-intestinal tract	1.1	1.7		13.4	14.8
Total	4.4	6.8		61.7	68.1
Skeletal muscle	27.0	41.5	0.5	13.5	14.9
Lung					
Skin	0.6	0.9	6.7	4.0	4.4
	5.0	7.7	0.3	1.5	1.7
Grand total	37.0	56.9		80.8	89.1

Note: Data for a 65-kg male with a BMR of 90.6 W

Source: Aiello, L. C.; Wheeler, P.: The Expensive Tissue Hypothesis. The Brain and the Digestive System in Human and Primate Evolution. Current Anthropology 36 (1995), pp. 199-221, 201.

Primary source of Aiello; Wheeler: Aschoff, J.; Guenther, B.; Kramer, K.: Energiehaushalt und Temperaturregulation. Munich 1971.

Figure 1: Relative brains size versus relative gut mass in primates.

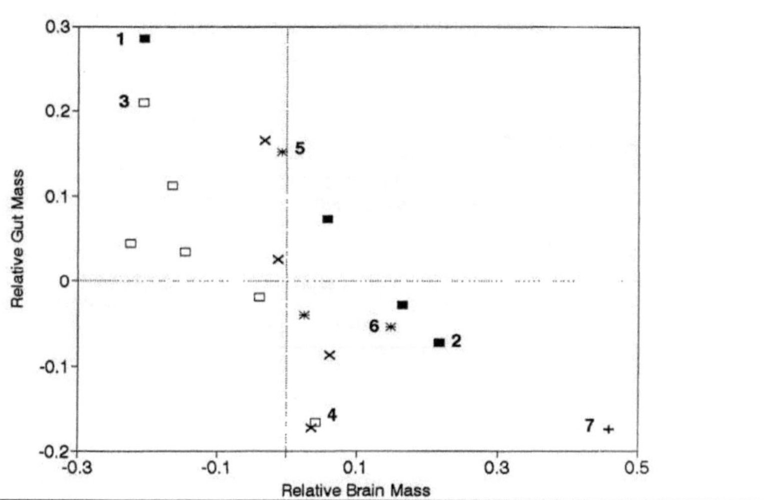

Note: The filled squares illustrate Cebids, while Colobines are shown by open squares. Stars ought to demonstrate the correlation of the brain size and gut mass of Hylobatids. X's illustrate other Catarrhines.

Besides:	1	Alouatta seniculus	5	Hylobates syndactylus
	2	Cebus apella	6	Hylobates lar
	3	Presbytis cristatus	7	Homo sapiens
	4	Presbytis rubicunda		

Source: Aiello, L. C.; Wheeler, P.: The Expensive Tissue Hypothesis. The Brain and the Digestive System in Human and Primate Evolution. Current Anthropology 36 (1995), pp. 199-221, 207.

Figure 2: Increased encephalization and high-quality diet based on animal protein.

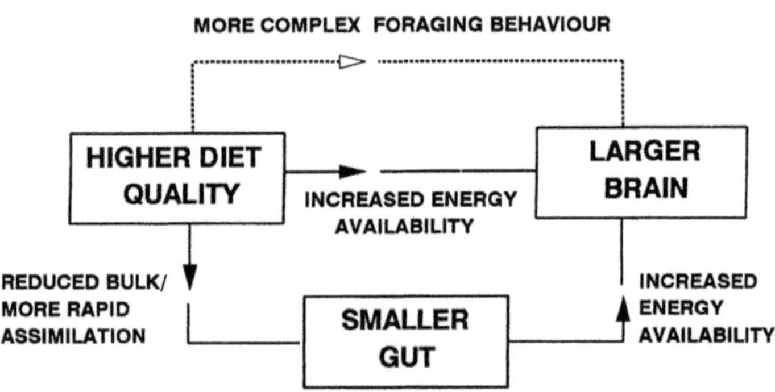

Note: The dashed lines illustrate a pressure that is selective while the solid lines indicate relaxed constraints.

Source: Aiello, L. C.; Wheeler, P.: The Expensive Tissue Hypothesis. The Brain and the Digestive System in Human and Primate Evolution. Current Anthropology 36 (1995), pp. 199-221, 207.

3. Bibliography

Aiello, L. C.; Dunbar, R. I. M.: Neocortex size, group size, and the evolution of language in the hominids. *Current Anthropology* 34 (1993), pp. 184-193.

Aiello, L. C.; Wheeler, P.: The Expensive Tissue Hypothesis. The Brain and the Digestive System in Human and Primate Evolution. *Current Anthropology* 36 (1995), pp. 199-221.

Aiello, L. C.: Brains and guts in human evolution. The expensive tissue hypothesis. *Brazilian Journal of Genetics* 20 (1997), pp. 141-148.

Aiello, L.C.; Key, C.: Energetic Consequences of Being a Homo erectus Female. *American Journal of Biology* 14 (2002), pp. 551-565.

Armstrong, E.; Bergeron, R.: Relative brain size in birds. *Brain, Behaviour, and Evolution* 26 (1985), pp. 141-153.

Aschoff, J.; Guenther, B.; Kramer, K.: Energiehaushalt und Temperaturregulation. Munich 1971.

Chivers, D. J.; Hladik, C. M.: Morphology of the gastrointestinal tract in Primates. Comparisons with other mammals in relation to diet. *Journal of Morphology* 166 (1980), pp. 337-386.

Chivers, D. J.; Hladik, C. M.: Diet and gut morphology in primates. In: Chivers, D. J.; Wood, B. A.; Bilsborough, A. (eds.): *Food acquisition and processing in primates*. New York 1984, pp. 213-230.

Conklin-Brittain, N. L.; Wrangham, R. W.; Smith C. C.: A Two-Stage Model of Increased Dietary Quality in Early Hominid Evolution: The Role of Fiber. In: Ungar, P. S.; Teaford, M. F. (eds.): *Human Diet. Its Origin and Evolution*. Westport 2002, pp. 61-76.

Dunbar, R. I. M.: The Social Brain Hypothesis. *Evolutionary Anthropology* 6 (1998), pp. 178-190.

Falk, D.: Brain evolution in Homo. The "radiator" theory. *Behavioral and Brain Sciences* 13 (1990), pp. 333-381.

Fish, J. L.; Lockwood, C.A.: Dietary Constraints on Encephalization in Primates. *American Journal of Physical Anthropology* 120 (2003), pp. 171-181.

Foley, R. A.; Lee, P. C.: Ecology and Energetics of Encephalization in hominid Evolution. *Philosophical Transactions of the Royal Society* 334 (1991), pp. 223-232.

Holloway, R. L.: Brain size, allometry, and reorganization. Toward a synthesis. In: Hahn, M. E.; Jensen, C.; Dudek, B. (eds.): *Development and evolution of brain size. Behavioral implications.* New York 1979, pp. 59-88.

Holloway, R. L.: Within-species brain-body weight variability. A reexamination of the Danish data and other primate species. *American Journal of Physical Anthropology* 53 (1980), pp. 109-121.

Kleiber, M.: The Fire of Life. New York 1961.

Leonard, W. R.; Robertson, M. L.: Evolutionary perspectives on human nutrition. The influence of brain and body size on diet and metabolism. *American Journal of Human Biology* 6 (1994), pp. 77-88.

Leonard, W. R.; Robertson, M. L.; Snodgrass, J. J.; Kuzawa, C. W.: Metabolic correlates of human brain evolution. *Comparative Biochemistry and Physiology Part A* 136 (2003), pp. 5-15.

Leonard, W. R.; Robertson, M. L.; Snodgrass, J. J.: Effects of Brain Evolution on Human Nutrition and Metabolism. *Annual Review of Nutrition* 27 (2007), pp. 311-327. In the following as Leonard; Robertson; Snodgrass (2007a).

Leonard, W. R.; Robertson, M. L.; Snodgrass, J. J.: Energetic Models of Human Nutritional Evolution. In: Ungar, P. S. (ed.): *Evolution of the Human Diet. The Known, the Unknown, and the Unknowable.* New York 2007, pp. 344-359. In the following as Leonard; Robertson; Snodgrass (2007b).

Mac Nab, B. K.; Eisenberg, J. F.: Brain Size and its Relation to the Rate of Metabolism in Mammals. *American Naturalist* 133 (1989), pp. 157-167.

Martin, R. D.: Primate Origins an Evolution. London 1990.

Milton, K: Primate diets and gut morphology. Implications for hominid evolution. In: Harris, M; Boss, E. B. (eds.): *Food and evolution. Toward a theory of human food habits.* Philadelphia 1987, pp. 96-116.

Ray, K.; Thomas, J.: In the kinship of cows. The social centrality of cattle in the earlier Neolithic of southern Britain. In: Parker Pearson, M. (ed.): *Food, Culture and Identity in the Neolithic and Early Bronze Age.* British Archaeological Reports International Series 1117 (2003), pp. 37-44.

Ruff, C. B.; Walker, A.: Body Size and Body Shape. In: Walker, A.; Leakey, R. E. (eds.): *The Nariokotome Homo erectus skeleton.* Cambridge 1993.

Schofield, W. N.: Predicting basal metabolic rate. New standards and review of previous work. *Human Nutrition: Clinical Nutrition Part C* 39 (1985), pp. 5-41.